Bioinformatics in Mass Spectrometry

Dr. Volker Egelhofer

Bioinformatics in Mass Spectrometry

A Practical Guide

*Bibliografische Information der Deutschen Nationalbibliothek:
Die Deutsche Nationalbibliothek verzeichnet diese Publikation
in der Deutschen Nationalbibliografie; detaillierte bibliografische Daten sind im Internet über http://dnb.dnb.de abrufbar.*

*© 2013 Dr. Volker Egelhofer
Herstellung und Verlag:
BoD – Books on Demand, Norderstedt*

ISBN: 978-3- 7357-9409-3

- Introduction 7
- Aim of the book 9
- Book organization 9
- Programming languages 10
  - Possible Classification 10
  - Source code 11
  - Interpreter 11
  - Compiler 12
    - Just-in-time compilation (JIT) 12
- The Python programming language 13
  - Operators 15
    - Comparison Operators 15
    - Logic Operators 16
    - Augmented Operators 16
  - Data types 17
    - Sequential data types 18
    - Strings 18
    - Lists 22
    - Dictionaries 25
    - Tuple 27
  - Files 28
  - Variables 30
  - Key words 30
  - Type checking 31
  - Indentation 32
  - Comments 33
  - Programming style 34
  - Exercise A1 35
  - Control Flow 37
    - Conditional statements 37
    - Loops 38
  - Exceptions 41
  - Exercise A2 43
  - Functions 45
    - Global and local Variables 47
  - Module 49

- Parameter passing 51
- Exercise A3 52
- Object-oriented programming OOP 54
  - Object 54
  - Class 55
  - Data abstraction 56
  - Inheritance 57
  - Encapsulation 58
  - Exercise A4 60
- Advanced Programming 61
  - Exercise B1 61
- Spectrum-to-Spectrum Search Algorithm 63
  - Protein identification algorithms 63
  - Experimental Steps 64
  - Exercise B2 65

Introduction

In the past, Bioinformatics has been mostly classified as a bridge discipline between Informatics and Biology rather than an independent (stand-alone) scientific discipline. But with the comprehensive accumulation of biological data and the resulting challenges Bioinformaticians concentrate more on their own research rather than simply serving as technologists for others. Nowadays, the focus in Bioinformatics is mainly on the development of sophisticated algorithms capable of extracting useful knowledge from large data sets by combining methods from statistics and artificial intelligence. Computational Proteomics and Metabolomics are the most emerging fields of bioinformatics. Metabolomics is the study of the small molecules (metabolites) in biological samples such as cells or tissues. This includes their identification and quantification as well as the interaction between them.

Proteomics is the study of the entire set of proteins of a given cell type, cell compartment or specific tissue under defined conditions at a specific time. Combined with high-resolution mass spectrometry (MS) the technology allows the identification and quantification of thousands of proteins. Protein separation by 2D- gel electrophoresis (2D-PAGE), followed by mass spectrometry (MS) or tandem mass spectrometry (MS/MS) identification is the classical method for protein identification by mass spectrometry. In the currently predominated shotgun proteomics strategy, a proteolytic digest of the protein sample is analysed by LC-MS/MS. In such a pipeline, one MS1 (full scan) spectrum is obtained roughly every second and a set of the most abundant ions from the MS1 scan are selected for fragmentation and recording of MS2 spectra (MS/MS).

The identification of peptides from acquired MS/MS spectra is either performed using a database search approach or a spectrum-spectrum search. In the first case the experimental m/z values are compared to calculated m/z values derived from peptides produced by an in silico digestion of a protein sequence library. This approach implies a robust and reliable functional protein sequence annotation. In the latter case the identification of peptides is carried out by matching spectra of unknown shotgun analyses against the reference spectra of the library. An advantages of that approach is a more simple visual discrimination of false positive identifications as well as a search time reduction due to smaller number of peptide sequence information.

***

Aim of the book

The aim of the book is to give the reader a basic insight of Mass spectrometry-based bioinformatics. This beginners guide will help to illustrate some of the common algorithmic problems occurring in typical high throughput mass spectrometry protein identification experiments. A general introduction to Python programming language including standard programming techniques and their role in problem solving will be provided.

***

Book organization

- Introduction to Python programming language
- Advanced programming techniques
- Development and Implementation of an algorithm for protein identification from sequence databases using mass spectrometry data.

***

Programming languages

A programming language is to be used for controlling the behavior of a computer and like human languages programming languages have syntactic and semantic rules used to define meaning.

Possible Classification

• Machine language
Computers only understand the so called machine language, which is composed of two symbols '0' and '1', or power 'on' – 'off'. It is not readable for humans.

• Assembly language
Low-level programming language. Each statement corresponds to a specific machine language instruction. Applied in electronic engineering programs where Code must interact directly with the hardware (for example in device drivers).

• High-level language
High-level languages are both better adapted to natural languages and portable to multiple systems (e.g.:BASIC, JAVA, C or C++).

• Fourth-generation language (4GL)
4GL languages are used for specific functions or complete applications to reduce programming effort. These includes SQL or R.

• Logic programming languages
These type of languages are used in the fields of artificial intelligence or computational linguistics. For example, Languages like LISP or PROLOG are used for programming expert systems, games, automated answering systems and other.

***

Source code

The source code also referred as "source" or "code" is the version of software as it is originally written in human readable plain text. To be usable by a computer the source code must be translated into machine language by special programs. There are two different approaches: Compiler translation and Interpreter translation.

***

Interpreter

In Interpreter is a computer program which translates text written in a computer language (the source code) into another lower level language, usually executable machine code which is output to a file for later execution.

Pros:
An advantage single is its platform independence as well as a line of code can be tested interactively.

Cons:
Interpreting code is more or less slow, because each statement in the source code must be analyzed each time it is executed.

Interpreter specific languages:
JavaScript, Lisp, Perl, PHP, Python, Ruby et al.

***

Compiler

A compiler is a computer program that translates the source code into another computer language (the object code). It performs many or all of the following operations: lexical analysis, pre-processing, parsing, semantic analysis, code optimization, and code generation. The name compiler is primarily used for programs that translate source code from a high level language to a lower level language (e.g., assembly language or machine language).

Pros:
A compiled program is fast, because it performs the desired action without time consuming translation steps.

Cons:
Program Development is more complex, harder to debug.

Compiler specific languages:
C, C++, C#, Ada, Pascal, Delphi et al.

Just-in-time compilation (JIT)

Another technique is Just-in-time compilation (JIT), which further blurs the distinction between interpreters, byte-code interpreters and compilation. JIT is available for example the .NET and Java platforms.

***

The Python programming language

Python is a portable, interpreted, object-oriented programming language. Its development started in 1990 at CWI in Amsterdam (Guido von Rossum) and continues under the ownership of the Python Software Foundation.

Important Features

Python is easy to use, no compilation step is needed, it is easy to debug, portable, open source and the language is widely spread in scientific environment.

Further information

- Newsgroup: comp.lang.python.*
- www: http://www.python.org.

Installation of the Python Interpreter

Note: The following is tested with python 2.7.5. Before continuing, please download the current version of Python (2.7.x) at http://www.python.org/download/ and follow the installation instructions.
If you have a computer with Mac OSX or Linux OS you may start the python run-time interpreter by opening a Terminal and enter 'python' into the window. On Windows computer please execute from the Run dialog box the program 'cmd.exe' to execute a terminal window. Now change the current directory in the terminal to the python installation directory and start the interpreter by entering 'python.exe'. The run-time interpreter is indicated by the prompt: ' >>>'. Any command that you enter here will be executed directly, variables are also stored (see below).

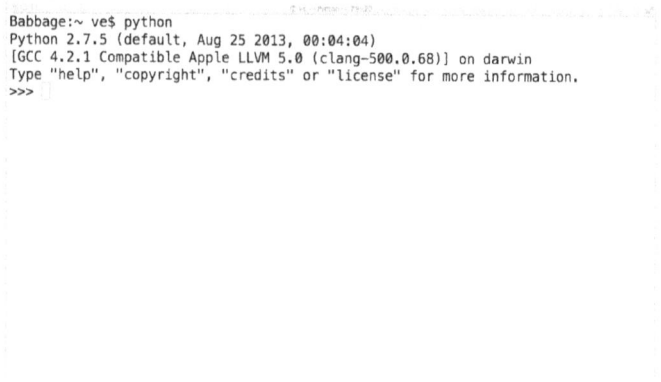

The key combination 'Ctrl + D' (Linux, OS X) or 'Ctrl + Z' (Windows) should terminate the interpreter, if not just type exit() within the terminal and hit <enter>. If you want to start a program directly with python, then you must pass the name of the script file as a parameter:

```
python myProgram.py
```

***

Operators

In computer science an 'operator' is a calculation rule. There are arithmetic, comparison and logical operators.

Arithmetic Operators

Operation	Operator	Example	Result
Addition	+	7 + 2	9
Subtraction	-	10-5	5
Multiplication	*	7 * 2	14
Division	/	7.0 / 2.0	3.5
Integer Division	//	7 // 2	3
Modulo Division	%	7 % 2	1
Exponentiation	**	7 ** 2	49
String concatenation	+	Pyth + "on"	Python

Comparison Operators

Comparisons are tested with the familiar < and > symbols used in mathematics for 'less than' and 'greater than'. 'Less than or equal to' and 'greater than or equal to' use the <= and >= operators respectively.

Comparison	Operator	Returns
Less than	<	True if left is less than right
Greater than	>	True if left is greater than right
Less than or equal	<=	True if left is not greater than right
Greater than or equal	>=	True if left is not less than right
Equal	==	True if left is equal to right
Not equal	!=	True if left is not equal to right

Logic Operators

Logical operators combine (usually two) logical expressions to one complex logical expression. The composite expression is either true or false (see below).

Logic	Operator	Returns
And	and	True only if left AND right are both true
Or	or	True if left OR right is true, or both are true
Not	not	True if right is false; False if right is true

Augmented Operators

Augmented assignment combines the = assignment operator with a binary arithmetic or bit-wise operator:

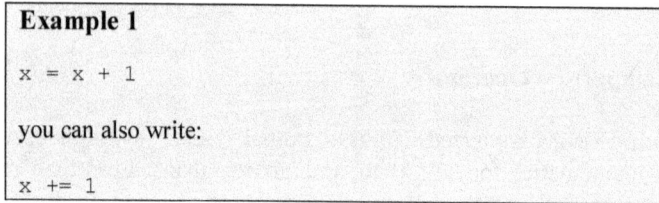

***

Data types

Each programming languages has its own set of predefined data types. These data types are used to fill allocated memory areas with concrete semantics (with data), which can be read out later. These storage areas are called 'variables' or 'constants'. The use of defined data types allows the compiler respectively the interpreter, to test the type compatibility of the operations specified by the programmer. Not allowed or not meaningful operations are thus detected and displayed to the programmer. The following provides a brief listing of some basic data types available in Python of particular use in scientific programming:

Integer

Integer are in the range of -2147483647 to 2147483647 for 32 bit systems.

Long Integer

Long Integer are integers too, but can be arbitrarily large, only restricted by the available memory of the computer. In Python, a long integer is declared by adding the letter "L" (allows lowercase or uppercase letters) to an Integer.

Floats

Floats are positive or negative real numbers. For larger values, for clarity, scientific notation is used, that is, we can insert the letter "e" or "E". Floating point numbers are stored internally only with finite precision and binary (i.e with the digits 0 and 1 instead of the digits 0 through 9). This results in rounding errors. Therefore, comparisons of floating point numbers are unreliable and should be avoided!

***

Sequential data types

Python provides so-called sequential data types: strings, lists, dictionaries and tuples.

Strings

A string may be empty, consists of several characters or of only one character. In a string can be included everything that can be converted into text. If you want to put control characters or other special characters in a string, you have to mask them with a backslash (\). A selection of these so-called *escape sequences* are shown in the following table.

Escape sequence	Meaning
\\	backslash
\'	single quote
\"	double quote
\n	newline(Linux)
\r\n	newline(Windows)
\r	newline(Mac)
\t	tab

The addition and multiplication of strings is allowed:

Example 2

```
>>> 'abc' + 'def'
'abcdef'

>>> 'ab' * 3
'ababab'
```

Strings may be indexed. An index starts with 0, two indices are separated by a colon!

Example 3
```
>>> str = 'example string'
>>> str[0]
'e'

>>> str[1]
'x'

>>> str[0:3]
'exa'

>>> str[2:7]
'ample'

>>> str[2:]
'ample string'
```

Strings must not be changed directly in Python, i.e. allocations to individual elements are not possible.

Example 4
```
>>> str = 'example string'
>>> str[0]='W'
```
Traceback (most recent call last):
File "<stdin>", line 1, in <module>
TypeError: 'str' object does not support item assignment

Special String Methods

Method	Description
count(*sub*, start, end)	Returns the number of occurrences of a substring *sub* in a string. Parameter 'start' and 'end' are optional.
startswith(*prefix*, start, end)	Returns True if a string starts with the *prefix*, otherwise returns False. Parameter 'start' and 'end' are optional.
endswith(*suffix*, start, end)	Returns True if a string ends with the specified *suffix*, otherwise returns False. Parameter 'start' and 'end' are optional.
find(*sub*, start, end)	Returns the lowest index in a string where a substring *sub* is found. Parameter 'start' and 'end' are optional.
split(*sep* ,maxsplit)	Returns a 'list' of words in a string, using *sep* as the delimiter string. The parameter 'maxsplit' is optional. If no parameter *sep* is given any white space character is used as delimiter.
strip(chars)	Returns a copy of a string with the leading and trailing control characters removed. Parameter 'chars' is optional, if set, the leading and trailing characters in 'chars' will be removed.
len(*s*)	Returns the number of characters in the string *s*.

Example 5

```
>>> str = 'an example'
>>> str.find('n')
1

>>> str.find('bei')
-1

>>> print s.count('e')
3

>>> print s.count('e',4)
9

>>> print s.count('e',4,7)
-1
```

Methods for type conversion

A selection of methods can be used for type conversion, are shown in the table below.

Method	Description
int(x)	converts x into an integer
long(x)	converts x in a long-integer
float(x)	converts x into a float
str(x)	converts x into a string

***

Lists

Lists consists of a number of objects which can be referenced by their position number within the object. In python, the list elements need not be of the same type. A list is defined with the square brackets ([]), the individual elements are separated by commas.

Example 6
```
>>> liste1 = ['Cat', 'small', 'tree']
>>> liste1[0]
'Cat'

>>> liste1[2]
'tree'
```

In contrast to strings, lists are modifiable:

Example 7
```
>>> liste1
['Cat', 'small', 'tree']

>>> liste1[0]='tiger'
>>> liste1
['tiger', 'small', 'tree']
```

Lists can be nested, i.e. a list may contain another list as an element.

Example 8

```
>>> liste1 = ['B', 'small', 'tree']
>>> liste2 = ['A', 'big', 'toe']
>>> liste3 = [liste1,liste2]
>>> liste3
[['B', 'small', 'tree'], ['A', 'big', 'toe']]

>>> liste3[0]
['B', 'small', 'tree']

>>> liste3[1]
['A', 'big', 'toe']

>>> liste3[0][1]
small
```

Special list methods

Method	Description
append(x)	Adds the element x to the end of the list.
extend(L)	Extends a given list by appending all the elements in the L.
insert(i, x)	Inserts the element x at the position i.
remove(x)	Removes the first element from the list whose value is equal to x.
pop(i)	Removes the element at position i in the list, and return it.
index(x)	Returns the index in the list of the first item whose value is equal to x.
count(x)	Returns the number of times x appears in the list.
sort()	Sorts the elements of the list, in place.
len(L)	Returns the number of elements in the list L.

Example 9

```
>>> list_A =[1,2,3]
>>> list_A.append(4)
>>> list_A
[1, 2, 3, 4]

>>> liste_B =[6,9,5]
>>> liste_B.extend(liste)
>>> liste_B
[6, 9, 5, 1, 2, 3, 4]

>>> liste_B.sort()
>>> liste_B
[1, 2, 3, 4, 5, 6, 9]
```

***

Dictionaries

It is best to think of a dictionary as an unordered collection of *key, value* pairs, with the requirement that the keys are unique (within one dictionary). The keys can be numerical constants, strings or other constant objects. The values could be of any kind of data type provided by Python. Dictionaries are defined by assigning its elements within curly brackets {}. Each element consists of a *key* followed by a colon and a *value*. Multiple items are separated by commas.

Example 10

```
>>> dictionary = {'jack': 4098, 'sparrow': 4139}
>>> dictionary['jack']
4098
```

The value of an element can be assigned via its key in square brackets.

Example 11

```
>>> dictionary['Will']=1234
>>> dictionary
{'Will': 1234, 'sparrow': 4139, 'jack': 4098}
```

Note: Keys must be unique (within one dictionary). Strings and numbers can be used as keys, but lists not, because they can be modified directly (e.g. with the append () method).

Example 12

```
>>> dictionary={}
>>> dictionary['jack']=1234
>>> dictionary['jack']
1234

>>> dictionary['jack']='different_Value '
>>> dictionary['jack']
'different_Value'
```

The *keys()* method of a dictionary object returns a list of all the keys used in the dictionary, in arbitrary order.

> **Example 13**
>
> ```
> >>> dictionary.keys()
> ['Will', 'sparrow', 'jack']
> ```

The *values()* method of a dictionary object returns a list of all the values used in the dictionary, in arbitrary order.

> **Example 14**
>
> ```
> >>> dictionary.values()
> [1234, 4139, 4098]
> ```

The 'has_key' method as well as the 'key in d' method can be used to test for the presence of key in the dictionary.

> **Example 15**
>
> ```
> >>> dictionary.has_key('jack')
> True
>
> >>> dictionary.has_key('jack')
> True
>
> >>> 'jack' in dictionary
> True
> ```

***

Tuple

A tuple is a special python type, it consists of a number of values separated by commas, similar to a list, but not they are not mutable. A Tuple is defined with the round brackets (()). Tuples have many uses. For example: (x, y) coordinate pairs, records from a database and other.

> **Example 16**
>
> ```
> >>> tuple = ('dog','cat')
> >>> tuple
> ```
> ('dog', 'cat')
>
> ```
> >>> tuple[0]
> ```
> dog
>
> ```
> >>> tuple[0]='horse'
> ```
> Traceback (most recent call last):
> File "<stdin>", line 1, in <module>
> TypeError: 'tuple' object does not support item assignment

Difference between Lists and Tuples

With a list, you can change the elements, add extra elements, and so on, after it has been created, with a Tuple not (see above).

***

Files

In Python files can be created or opened with the method: *'open(filename, mode)'*, where the parameter *'mode'* refer to one or more of the following characters:

mode	Description
r	read
w	write (overwrite)
a	append

The following command is used to create a file named 'workfile' in the directory '/tmp/workfile':

Example 17

```
>>> f=open('/tmp/workfile', 'w')
```

Note: If a file with the same name already exists, it is opened, but its contents will be overwritten.

The following function is used to write a string to the file. There is no return value.

Example 18

```
>>> f.write('This is a test\n')
```

After all operations the file should be closed. Calling *close()* more than once is allowed.

Example 19

```
>>> f.close()
```

Special File Methods

Example 20
```
# first create a new example file
>>> f = open('workfile.txt', 'w')
>>> f.write('0123456789abcdef')
>>> f.close()

# open the new file for reading, the File-cursor is then at position '0'
>>> f = open('workfile.txt', 'r')

# read one Byte(here equal to a character) from the file
>>> f.read(1)

# place the File-cursor at position '5'
>>> f.seek(5)

# read again one Byte)from the file
>>> f.read(1)

# close the file
>>> f.close()
```

***

Variables

The computer uses numerical addresses to manage its memory. Memory addresses are fixed-length sequences of bits. Variables are names for certain areas of this memory addresses, which could be used to modify the content of certain address.

> **Example 21**
> \# *s* is a string variable
> ```
> >>> s = ' the value of the variable s '
> ```
> \# *i* is an integer variable
> ```
> >>> i =0
> ```

***

Key words

Key words are reserved words which must not be used as names for variables, functions or classes!

Python provides the following key words:

and ,assert ,break ,class ,continue ,def ,del ,elif ,else ,except ,exec ,finally ,for ,from ,global ,if ,import ,in ,is ,lambda ,not ,or ,pass ,print ,raise ,return ,try ,while ,yield

***

Type checking

var *x* := 7; (1)
var *y* := "hallo"; (2)
var *z* := x+y; (3)

1. binds the value 7 to x;
2. binds the value "hallo" to y; and
3. attempts to add x to y.

Note: In a strong-typed language, the value bound to *x* might be a pair (integer, 7) and the value bound to *y* might be a pair (string, "hallo"). When the program attempts to execute line 3, the language implementation would check the type tags integer and string, discover that the operation + (addition) is not defined over these two types, and an error would be indicated!!! A weakly-typed language would produce:" 7hallo"

***

Indentation

One unusual aspect of Python's syntax is its use of white space or tabs to delimit program blocks (the off-side rule). In so-called "free-format" languages, that use the block structure ultimately derived from ALGOL, blocks of code are set off with curled braces { } or keywords. In these languages, however, programmers conventionally indent the code within a block, to set it off visually from the surrounding code. Python, instead, borrows a feature from the lesser-known language ABC -instead of punctuation or keywords, it uses this indentation itself to indicate the run of a block.

Note: Better do not mix tab and space characters. The indentation length is not the length you see in the buffer, but equal to the number of separation characters!

Please note the following rules:

i. The first line of a program must not start with a space or a tab!
ii. A colon and an indentation in the next line is always necessary when a block is opened.
iii. All lines that belong to the same block must be indented to the same column.
iv. Everything that is indented at one level, belongs to the same block.

Example 22
```
>>> s = 'string'
>>> s
```
'string'

```
>>> s
```
File "<stdin>", line 1
 s
 ^
IndentationError: unexpected indent

Comments

In Python comments are marked with a #, that is, everything written right of this character until the end of the line is treated as a comment and will not be executed. Whole blocks can be commented out with three " at the beginning and end.

> **Example 23**
> \# this is a comment
>
> """ the comment reaches from here
> MAX = 19
> values_10_3 = [1, 10, 30, 32, 18, 19, 33, 7, 11, 14, 16, 2]
> values_10_4 = [1, 10, 30, 32, 18, 19, 33, 7, 11, 14, 35, 16, 2, 0]
> """ to there

Note: Code itself should be self documenting: use descriptive class, member, variable, parameter and method names. You should also comment each code block of your program, using a uniform approach for each level.

***

Programming style

The names of classes, functions or variables should be clear and self explanatory.

$\Rightarrow$ use 'faculty' instead of 'f'

Single letters such as i or j should only be used as control variable of loops or as parameters of simple mathematical functions e.g. x=x+2

$\Rightarrow$ use 'linePerPage' instead of 'z'

Use case-sensitive Characters and/or underscores to separate parts of words
$\Rightarrow$ use 'calculateFaculty' instead of 'calculate_Faculty'

***

Exercise A1

1. Use the Python interpreter. Enter the following commands, jot down the result and discuss it.
a. 3==4
b. 3.0==3
c. int('3')
d. int(3.9)
e. int('3.9')
f. float(3.9)
g. 7/2
h. 7/2.0
i. 'aaa' + 5
Stop the Interpreter by pressing Ctrl+D (Linux,Mac), Ctrl + Z (Windows).

2. Use a text editor. Write a program to print the text 'Hello World' on the screen. Save the file with the extension '.py' in your python installation directory. To test your Code, open a shell, change to the python installation directory and enter 'python Name_of_your_script.py' (OSX,Linux) or 'python,exe Name_of_your_script.py' (Windows) .

3. Given a string str = 'This is a string'. Write a program, that
a. prints out the second character of str
b. replaces the substring 'is' in str with 'is not' and prints out the result
c. splits str into words, and prints out the resulting list
on the screen.

4. Given a list L = ["john","tim","chris","mike"]. Write a program, that prints out
a. the entire list
b. the first three elements of the list L
c. the second and third element of the list L
on the screen.

5. Learn more about the functions 'input' and 'raw_input' at http://docs.python.org/2/library/functions.html. Write a pro-

gram, where the user is asked to enter his/her first and last name. The program should print out the user input provided with a welcome line on the screen.

6. Write a program, where the user can enter a number. The program should notify the user if the entered number is even or odd.

7. Write a program, that asks the user to enter the names of three Nobel Laureates and the year the prizes was awarded. Save the input data in a list.

8. Expand the program of exercise 7. The program should now write the entries line by line (eg 'name' ,'year ') to a file.

9. Write a program which reads the file from exercise 8 and prints out its content in tabular form (use '\t')and in reverse order to the screen.

***

Control Flow

The main structure of a computer program is sequential. Statements are executed in sequence until an instruction or statement changes the flow of control. Python provides control structures that serve to specify what has to be done by your program, when and under which circumstances.

Conditional statements

One of the most important control structures are conditional statements, they enable a program to respond to different states and inputs.

IF <condition> is TRUE
THEN:
<do something>
ELSE:
<do something else>

in python:

Example 24
```
if number<0:
  print 'number is less than zero'
elif number==0:
  print "number is zero'
else:
  print "number is greater than zero'
```

***

Loops

Loops have as purpose to repeat a statement a certain number of times or while a condition is fulfilled.

<u>while-loop</u>

The statements inside the loop are executed as long as a certain condition is true. While loops should be used, if you do not sure how many iteration steps are required to reach the end of the loop.

while condition == TRUE:
statement(s) ...

Example 25

```
>>> n=0
>>> while n<3:
...print n
...n+=1
...
```

<u>Output</u>:

0
1
2

***

break

Using *break* we can leave a loop even if the condition for its end is not fulfilled. It can be used to end an infinite loop, or to force it to end before its natural end.

while 1:
do something()
if condition == True:
break

Example 26

```
>>> n=0
>>> while 1: # this condition is always true!
...print n
...if n>1:
... break
...n+=1
```

Output:

0
1
2

***

for-loop

Its main function is to repeat statement while condition remains true, like the while loop. But in addition, the for loop provides specific locations to contain an initialization statement and an increase statement. So this loop is specially designed to perform a repetitive action with a counter which is initialized and increased on each iteration.

for i in list:
print i

in detail:

1. Initialization is executed. Generally it is an initial value setting for a counter variable (i=0). This is executed only once.

2. Condition is checked. If it is true the loop continues, otherwise the loop ends. (after last element in the list is reached)

3. Statement is executed. (print i)

4. Finally, loop gets back to step 2.

Example 27

```
>>> list = [1,8,124,3456]
>>> for i in list:
...print i
...
```

Output:
1
8
124
3456

***

range

The range function creates a list containing arithmetic progressions. It is most often used in for loops. The arguments must be plain integers. If the step argument is omitted, it defaults to 1. If the start argument is omitted, it defaults to 0.

Example 28

```
>>> for i in range(3,11,2):
... print i
...
```

Output:
3
5
7
9

Example 29

```
>>> for i in range(3,7):
...print i
...
```

Output:
3
4
5
6

Example 30

```
>>> for i in range(3):
... print i
...
```

Output:
0
1
2

***

Exceptions

It is possible to write programs that handle selected exceptions:

```
try:
    statements 1
    statements 2
    statements 3
...
except xyz-error:
... print xyz-error
... # handle exception
```

in detail:

First, the so-called 'try clause' - the statement(s) between the try and except keywords- is executed. If no exception occurs, the except clause is skipped and the execution of the try statement is terminated. If an exception occurs during the execution of a statement within the try clause, the follow up statements are skipped. The next step is comparing the type of the exception with the exception named after the except keyword. If they match, the so-called 'except clause' is executed, and then execution continues after the try statement. If an exception occurs which does not match the exception named in the except clause, it is passed either on to outer try statements (not shown) or if no handler is found, it is an unhandled exception and execution stops.

> **Example 31**
>
> ```
> >>> list = [22, 2.5, 0, -1]
> >>> for n in list:
> ... print n,
> ... try:
> ...   print 1 / n
> ... except ZeroDivisionError:
> ...   print 'an error occurred'
> ```
>
> Results:
> 22 0
> 2.5 0.4
> 0 an error occurred
> -1 -1

The *try ... except* statement has an optional else clause, which, when present, must follow all except clauses.

> **Example 32**
>
> ```
> >>> for arg in sys.argv[1:]:
> ...   try:
> ...     f = open(arg, 'r')
> ...   except IOError:
> ...     print 'cannot open', arg
> ...   else:
> ...     print arg, 'has been opened '
> ...     f.close()
> ```

***

Exercise A2

1. Write a program, that counts both the number of lines and number of words in a file and prints the results to the screen.

2. Write a program, that asks the user to guess a number between 1 and 10. The maximum allowed attempts are 10. (To generate a random number please use the module 'random' from the Python library).

3. Write a program, that calculates the horizontal cross sum of a number entered by the user and print it to the screen.

4. Write a program, where the user can enter an amino acid (one-letter code). The program should determine the chemical properties of the amino acid and print it out to the screen. It should be possible to distinguish between the following properties: hydrophobic, hydrophilic, basic, acidic. In addition, the program should catch incorrect inputs and exit when a blank character is entered.

5. Write a program, where the user can enter a string. The program should determine whether the user-entered string is a

nucleotide sequence, an amino acid sequence or neither and prints out the result to the screen.

6. Extend your program from exercise 5, to determine and display the number of different nucleotides if a nucleotide sequence was entered respectively the number of different amino acids when an amino acid sequence was entered. Furthermore, the program should catch incorrect inputs and exit when you enter a blank character.

Advanced Exercises

I. Write a program, that simulates the growth of a bacterial culture in an optimum environment within 2 days. The initial number of cells should be 100 and the population is doubled every 30 minutes.

II. Enhance your bacterial growth simulation such that the growth rate depends on the current size of the population. The growth rate should be $1+((K-p)/K)$ where K is the maximum population capacity and p is the current size of the population. Write a function that returns the growth rate in dependence of the capacity and population. (K = 10000000).

***

Functions

A function is a piece of code that performs a specific sub-task and is defined using the def keyword followed by the name of the function, a pair of parentheses (could be empty) containing a list of parameters, and a colon.

def func_name(param1, param2, ...):
 statement(s)
 return value # optional

Example 33

```
>>> def myFunction(): # Function declaration
... for i in range(1,5):
...print i
...
...# now call the function called 'myFunction' declared above

>>> myFunction()
```

Output:

1
2
3
4

It's possible to use default values. These could be assigned to their parameter names by using an equal sign.

def myfunction(value=4, string='value unequal 4'):
 if value != 4:
  print string
 else:
  print 'ok'

Arguments with assigned default values need not be specified when calling the function. All of the following calls are permitted:

myfunction()

myfunction(5)

myfunction(5,'new')

Python provides a large number of useful, predefined functions, which makes the work of developers easier, e.g.: the input () function. Each function has a specific task. In the case of the function input (), this is to accept an input.

Example 34

```
>>> z = input('Please enter a number: ')
>>> print 'the entered number is =', z
```

Output:

the entered number is = 345

***

Global and local Variables

All Variables have a type but are never declared in Python. They are instantiated when they are assigned for the first time. By default, variables are defined in the local name-space, or have to be declared explicitly as global variables, using the global statement.

Caution
The first assignment of a value stands for the variable declaration. If a value is assigned to a variable in a function body, the variable will be local, even if there is a global variable with the same name, and this global variable has been used before the assignment.

Global Variables

These are defined at the level of a module and available everywhere in the entire program where the respective modules are integrated.

Local Variables

These are defined within a function. They are then only available within the respective function and exist only as long as the function is active.

Example 35

```
def setValue(a):
 x=a+3 # local definition of x
 print x # print out the value of x
```

Call:

```
x=5 # (1)
print x # (2)
setValue(x) #(3)
print x # (4)
```

(1) global definition of x
(2) print out the value of the global variable x
(3) call the function 'setValue'
(4) print out the value of the global variable x, after calling the function 'setValue'

***

Module

A module is a collection of related program statements, mostly function definitions, which can be used by importing the module in other programs.

Note: A module is loaded only once, i.e, a second import statement will neither re-execute the code inside the module.

Be aware of potential names collision: For instance, if you current namespace contains a definition of a variable called, say: count, it will be destroyed and overloaded by the string module's definition of the count function.

Example 36

```
# mymodule.py: module description

def calcSum(a,b):
    return a+b

def calcProduct(a,b):
    return a*b
```

Note: To make use of the functions defined in the module 'mymodule.py', you must import the module. This could be done with the command 'import'.

```
import mymodule (1)
a = 2
b = 4
sum = mymodule. calcSum(a,b) (2)
product = mymodule. calcProduct(a,b) (3)
print sum,product (4)
```

(1)# imports the module mymodule.py
(2) Call the Function 'calcSum' defined in 'mymodule.py'
(3) Call the Function 'calcProduct' defined in 'mymodule.py'
(4) prints the results of (2) and (3)

Python provides the programmer with a large number of standard modules.

Selected standard modules

Modul	Description
sys	Interaction with the interpreter
exit()	The program will be terminated immediately.
argv	List of command line parameters which will be passed to the program being launched.
path	provides access to the system path
string	Allows for the manipulation of strings.
split	splits Strings in sub-strings and stores them in a list.
upper	capitalization.
lower	lowercase.
os	allows you to interact with the operating system.
open	opens a file.
system	executes system commands.
mkdir	creates a directory.
getcwd	returns the path name of the current working directory.
re	this module provides regular expression matching operations similar to those found in Perl.
search	finds particular patterns in a string.
subn	replacement of strings.
math	provides access to the mathematical functions defined by the C standard.
sin, cos	trigonometric Functions.

***

Parameter passing

In Python parameter passing is always done via 'Call_By_Reference', e.g. if calling a function with parameters, the references (memory addresses) of the parameters are passed to the function and not their values. Thus, the parameters passed into the function can be changed permanently. Exceptions are simple data types such as integer, float and strings. These data types are passed via 'Call_By_Value', which means there is a copy of them passed and not their memory addresses. Thus, the so passed parameters can not be changed permanently.

> **Example 37**
>
> ```
> def setAndPrintX(x=[]): (1)
>  x[0]=5 (2)
>  print x[0] (3)
>
> def setAndPrintY(y=0): (4)
>  y=5 (5)
>  print y (6)
> ```
>
> (1) x is a list, therefore passed via Call_By_Reference
> (2) x at position 0 will be changed permanently
> (3) print the first entry of the list x
> (4) x is an Integer, therefore passed via Call_By_Value
> (5) y is only changed locally not permanently
> (6) print y
>
> Call functions:
>
> ```
> x=[1,2,3,4]
> y=1
> print x[0] (7)
> setAndPrintX(x)
> print x[0]
> print y
> setAndPrintY(y)
> print y
> ```

> Results:
>
> 1 # before calling the function 'setAndPrintX'
> 5 # in function 'setAndPrintX'
> 5 # after calling function 'setAndPrintX'
> 1 # before calling function 'setAndPrintY'
> 5 # in function 'setAndPrintY'
> 1 # after calling function 'setAndPrintY'

Exercise A3

1. Write a function 'diff', that compares two lists and returns true if the lists are equal and false otherwise.

2. Write a function 'maximum', which calculates and returns the maximum of two numbers. Write a second function that calculates and returns the maximum of three numbers. Test both functions.

3. Write a function, that returns the largest element of a given list of integers.

4. Write a function, that both capitalize the first letter of each line in a text file, and numbers the rows in ascending order. The result should be printed to the screen. In addition, the path of text file should be passed to the program being launched.

5. Write a function, which expects a string as parameter. The function should check, whether the string is a palindrome and if so return True, otherwise False. Test your function.

Advanced Exercises

I. Write a function, that calculates the factorial of a natural number and use this function in a Python script in order to calculate n! for n = 1 to 10. The script should print a tabular output on the screen.

II. Write a function, that determines whether a particular sequence pattern exists in a given DNA sequence.

III. Extend your program from exercise 8, to determine the positions where the sequence pattern was found in the sequence and display number on the screen.

***

Object-oriented programming OOP

The basic idea of this concept is to integrate coherent data and functions or methods that can be applied to these data in a so-called object. Access to the data (attributes) of an object is done exclusively via the call to the object associated functions (methods). The goal is a strict separation of functions and data. An object-oriented program can therefore be considered as a system of communicating objects. Object-oriented programming provides greater flexibility and maintainability in programming, it is widely popular in large-scale software engineering.

Procedural vs Object-oriented (OO) programming

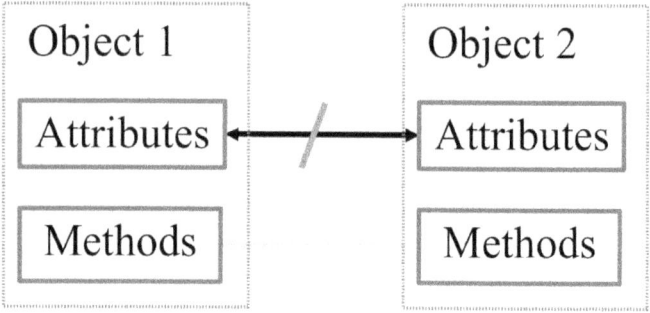

Object

An Object is the run-time manifestation (instantiation) of a particular exemplar of a class. Each object has its own data, though the code within a class (or a subclass or an object) may be shared for economy. Thus, object-oriented languages must allow code to be reentrant.

***

Class

The class can be regarded as a building plan for the data and their behavior (functionality). The class is the basic structure in an object-oriented computer program. A class should be typically recognizable to a non-programmer who is familiar with the problem domain. The code for a class should be (relatively) private and independent.

***

Features of OOP

- Data abstraction
- Data encapsulation
- Inheritance
- Polymorphisms

Advantages of OOP

- high reliability
- low amount of maintenance
- better reusability

Data abstraction

| real world | *airplanes* |

| data abstraction | *class* | **Properties** |
|||year of construction|

Properties
- year of construction
- range
- status

Methods
- start()
- fly()
- land()

| instantiation | *Object* |

Object 1

year of construction = 2001
range = 7000
status = parking
….

Object 2

year of construction = 1996
range = 12000
status = landing
…

***

Inheritance

Classes can inherit from other classes.

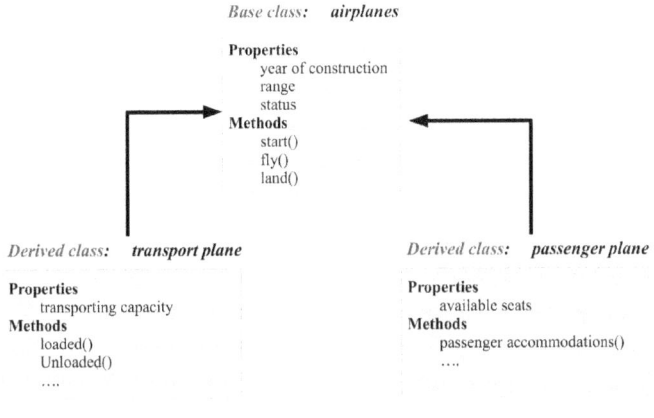

***

Encapsulation

A direct access to the internal properties of a class is not allowed, instead it is carried out via defined interfaces (methods of the class)

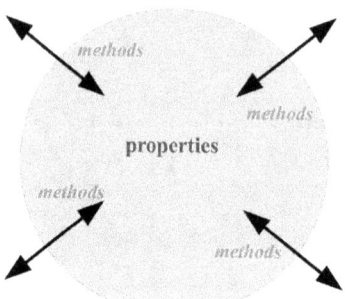

Implementation in Python

Example 38

```
class NameOfClass: (1)
 self.anAttributeOfClass='?' (2)

 def __init__(self): (3)
  pass

 def methodsOfClass(self): (4)
  print self.text
```
(1) name of the class
(2) attribute of this class
(3) Constructor
(4) additional methods

Note: The method '__init__' of a class is called *constructor*. The so-called '*constructor*' method of a class is always called first whenever an instance (object) of the class is created.. All methods defined within a class have a 'self' as the first parameter.

In Python instances of classes are created as follows:

> **Example 39**
>
> ```
> class PrintText:
>   def __init__(self, aString):
>     self.str = aString
>
>   def printOut(self):
>     print self.str
>
> printTextObject = PrintText('ExampleText')(1)
> printTextObject.printOut ()(2)
> ```
>
> Output:
>
> 'ExampleText'
>
> *(1) While executing that command, an object of the class 'PrintText' called 'printTextObject' is created. The passed string 'ExampleText', is then stored in the variable self.str.*
>
> *(2) By calling the method 'printOut' of the class 'PrintText' the value of the variable self.str will be printed to the screen.*

Note: Using Python it is neither necessary nor possible to delete objects. Python automatically takes care of objects that have no more active references.

***

Exercise A4

1. Define a class named 'BioSequence', which represents the most important features of protein sequences. Thus it should be possible both 'write' and 'read' Protein-Accession numbers, protein names, sequence lengths. A parameter should be passed the constructor at initialization.

2. Define two other classes named 'Protein' and 'Peptide', which are derived from the base class 'BioSequence'. The classes should have their own constructor, which expects a sequence as parameter. Furthermore, the constructor should check the correctness of the sequence.

3. Add a method named 'digest' to the protein class. The method should allow for the in silico digestion (using trypsin) of the given sequence. As result it should return a list of all digested peptides. Check this method with an example sequence.

4. Add another method called 'getMolecularWeight' to the protein class, which calculates the molecular weight of an amino acid sequence in Dalton and returns the result. Test this method too.

5. Write a program, that allows you to test all classes defined in the above exercises (1-4).

***

Advanced Programming

The most common text-based format for representing nucleic acid and protein sequences is the FASTA format (see example below). The amino acids are represented by their one-letter code. Each entry begins with the larger symbol '>' followed by the title, which includes descriptions of the sequence (usually protein name, database accession number, etc.), or comments. The sequence starts directly at the next line. (Usually each line should contain less than 60 characters). A file can contain any number of entries.

Example 40

```
>gi|532319|pir|TVFV2E|TVFV2E envelope protein
ELRLRYCAPAGFALLKCNDADYDGFKTNCSNVSVVHCTNLMNTTVTTGLLLNGSYSENRT
QIWQKHRTSNDSALILLNKHYNLTVTCKRPGNKTVLPVTIMAGLVFHSQKYNLRLRQAWC
HFPSNWKGAWKEVKEEIVNLPKERYRGTNDPKRIFFQRQWGDPETANLWFNCHGEFFYCK
MDWFLNYLNNLTVDADHNECKNTSGTKSGNKRAPGPCVQRTYVACHIRSVIIWLETISKK
TYAPPREGHLECTSTVTGMTVELNYIPKNRTNVTLSPQIESIWAAELDRYKLVEITPIGF
APTEVRRYTGGHERQKRVPFVXXXXXXXXXXXXXXXXXXXXXXXVQSQHLLAGILQQQKNL
LAAVEAQQQMLKLTIWGVK
```

***

Exercise B1

1. Download the sample file "fasta_example.zip" at "http://www.scienceread.com/".

2. Write a program, which prints out the first line of the file 'fasta_example.txt' on the screen. It should be possible to pass the file name as parameter, when you run your program. If the file is not found or can't be read, the program should catch an 'IOError' thrown by the interpreter.

3. Write a program that displays the title of all entries on the screen. How many entries are in the file?

4. Write a program that prints out the names and accession numbers of each entry (comma separated) on the screen.

5. Extend your program of exercise 3. It should print out the corresponding sequence too. Define a class 'Fasta', which represents the contents of fasta files (title + sequence). It should be possible to assign the title line, the sequence and the length of the sequence and read it again.

6. Write a program that outputs all entries originates from both 'EMBL' and 'GENBANK' database. How many entries are from the 'PDB' database?

7. How many entries has a sequence with length > 200?

8. What is the accession number of the entry with the shortest and longest sequence? Is there more than one entry?

***

Spectrum-to-Spectrum Search Algorithm

It is possible to draw conclusions about common functions of Proteins based on a sequence similarity. This is based on the assumptions that a function of a protein is determined by its amino acid sequence and the probability that two similar sequences could be caused by chance is very small.

Biological databases

Biological databases can be divided into sequence and functional databases. Sequence databases contain information of nucleotide sequences of genes or information about amino acid sequences of proteins. Functional databases contain additional information, such as function, structure and similarity of biological sequences.

Protein identification algorithms

Type I: Sequence Database depended identification

Algorithms match the experimental obtained fragment spectra to in-silico calculated peptide fragment mass spectra extracted from annotated protein sequence databases. A score is assigned to each candidate peptide. Depending on the size of the database the number of candidate peptides ranges from hundreds to hundreds of thousands. Database search methods differ primarily in their choice of score function.

Type II: Spectrum-To-Spectrum identification

Identify spectra by comparing them to a library of previously identified spectra. pros: searching an observed spectrum against real spectra is likely to give better results than searching against theoretical spectra. cons: library may contain false positives spectra.

Experimental Steps

I. The proteins are digested into peptides using trypsin.

II. The peptides are then separated by liquid chromatography, to reduces the complexity of the mixtures of peptides going into the mass spectrometer.

III. The mass spectrometer analyzes about 20,000 peptides - exited from the LC- per second.

IV. The ms selects a number distinct peptide species (Precursor ions) for fragmentation.

V. Each of these species is isolated and subjected to a second round of mass spectrometry analysis.
Beside the m/z-value as well as the intensity-value of the precursor ion, the resulting 'fragmentation spectra' are the third important output of the experiment.

Challenges:

I. Expected fragment ions will fail to be observed.

II. Additional unknown peaks -because of unusual fragmentation events or contamination- will be observed.

Exercise B2

Develop a spectrum-to-spectrum search algorithm for the identification of peptide mass spectra.

1. Download the sample file 'samples.zip' containing fragment ion spectra in the MGF-format at "http://www.scienceread.com/".

2. Write a parser, that extracts both m/z and the charge of the precursor, as well as all m/z – and intensity values of the fragment ions of each entry and store them in an appropriate format.

3. Download the library file 'database.zip containing fragment ion spectra in the MGF-format at "http://www.scienceread.com/".

4. Write a parser, that allows you to extract all relevant information from the library (i.e. Scan number, precursor m/z, precursor charge, m/z and intensities of the fragment ions) and store them in an appropriate format.

5. Develop an algorithm, that allows for the identification of the unknown spectra using information stored in the library.

General Procedure:

Analysis of the project:

- Working steps
- Which methods are required?
- How do I evaluate (a scoring algorithm) the results?
- Documentation of all methods

***

www.ingramcontent.com/pod-product-compliance
Lightning Source LLC
Chambersburg PA
CBHW050239230526
45470CB00005B/2030